Let's Read About Food

Meat

by Cynthia Klingel and Robert B. Noyed
photographs by Gregg Andersen

Reading consultant: Cecilia Minden-Cupp, Ph.D.,
Adjunct Professor, College of Continuing and Professional Studies, University of Virginia

For a free color catalog describing
Weekly Reader® Early Learning Library's
list of high-quality books, call 1-800-542-2595
or fax your request to (414) 332-3567.

Library of Congress Cataloging-in-Publication Data available
upon request from publisher. Fax (414) 336-0157 for the
attention of the Publishing Records Department.

ISBN 0-8368-3058-X (lib. bdg.)
ISBN 0-8368-3147-0 (softcover)

This edition first published in 2002 by
Weekly Reader® Early Learning Library
330 West Olive Street, Suite 100
Milwaukee, WI 53212 USA

Copyright © 2002 by Weekly Reader® Early Learning Library

An Editorial Directions book
Editors: E. Russell Primm and Emily Dolbear
Art direction, design, and page production: The Design Lab
Photographer: Gregg Andersen
Weekly Reader® Early Learning Library art direction: Tammy Gruenewald
Weekly Reader® Early Learning Library production: Susan Ashley

All rights reserved. No part of this book may be reproduced, stored in a retrieval system, or transmitted in any form or by any means, electronic, mechanical, photocopying, recording, or otherwise without the prior written permission of the copyright holder.

Printed in the United States of America

1 2 3 4 5 6 7 8 9 06 05 04 03 02

Note to Educators and Parents

As a Reading Specialist I know that books for young children should engage their interest, impart useful information, and motivate them to want to learn more.

Let's Read About Food is a new series of books designed to help children understand the value of good nutrition and eating to stay healthy.

A young child's active mind is engaged by the carefully chosen subjects. The imaginative text works to build young vocabularies. The short, repetitive sentences help children stay focused as they develop their own relationship with reading. The bright, colorful photographs of children enjoying good nutrition habits complement the text with their simplicity and both entertain and encourage young children to want to learn — and read — more.

These books are designed to be used by adults as "read-to" books to share with children to encourage early literacy in the home, school, and library. They are also suitable for more advanced young readers to enjoy on their own.

— Cecilia Minden-Cupp, Ph.D.,
Adjunct Professor, College of Continuing and Professional Studies, University of Virginia

I like to eat meat.
It is good for me.

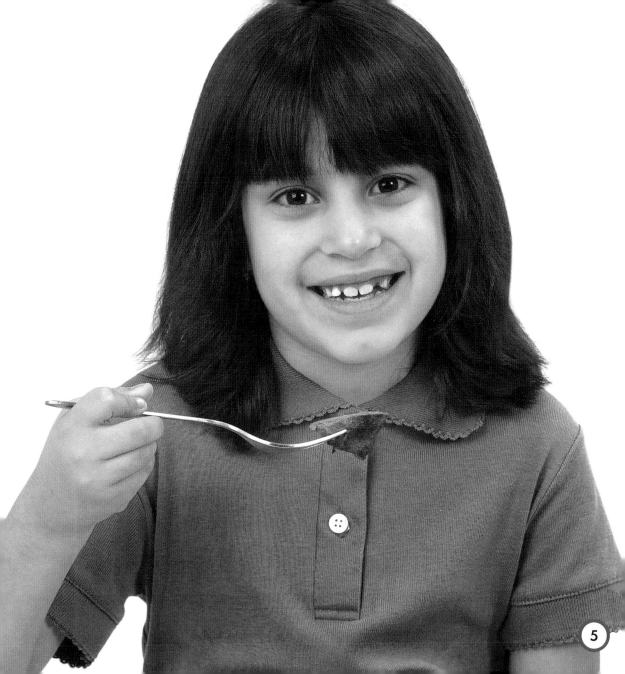

We choose from six different kinds of food. We need to eat all six kinds every day to stay healthy.

My body gets protein, calcium, and iron from meat. These make me strong.

Beef, pork, and lamb are kinds of meat. So are chicken, turkey, and fish!

Chicken is my favorite meat.
I like the wings!

Hot dogs are made of meat.
I want a hot dog at the game.

I like different kinds of fish, too. Tuna fish and fish sticks are yummy.

This group of food includes eggs, beans, and nuts.
I like to eat peanuts for a snack.

Right now, I am hungry for a hamburger!

Glossary

calcium—a soft, silvery-white element found in teeth and bones

iron—an element found in some foods

protein—an element found in all living plant and animal cells that is necessary for all life

For More Information

Fiction Books

Jaffe, Nina, and Louise August. *The Way Meat Loves Salt.* New York: St. Martin's Press, 1998.

Reid, Suzan, and Linda Hendry. *The Meat Eaters Arrive.* New York: Firefly Books, 1996.

Nonfiction Books

Clark, Elizabeth A., and John Yates. *Meat.* Minneapolis, Minn.: Lerner, 1990.

Frost, Helen, and Gail Saunders-Smith. *The Meat and Protein Group.* Mankato, Minn.: Pebble Books, 2000.

Web Sites

Meals.Com
www.my-meals.com/KidsRecipes.asp
For some easy recipes for kids

Index

beans, 18
beef, 10
calcium, 8
chicken, 10, 12
eggs, 18
fish, 10, 16
hot dogs, 14

iron, 8
lamb, 10
nuts, 18
pork, 10
protein, 8
turkey, 10

About the Authors

Cynthia Klingel has worked as a high school English teacher and an elementary school teacher. She is currently the curriculum director for a Minnesota school district. Cynthia Klingel lives with her family in Mankato, Minnesota.

Robert B. Noyed started his career as a newspaper reporter. Since then, he has worked in school communications and public relations at the state and national level. Robert B. Noyed lives with his family in Brooklyn Center, Minnesota.